I0030807

SUPER SIMPLE Science EXPERIMENTS

Laboratory Notebook

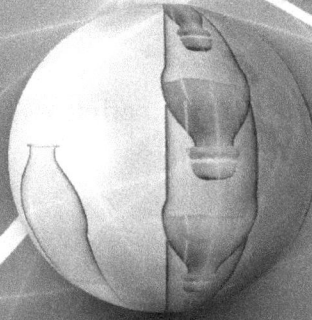

Name

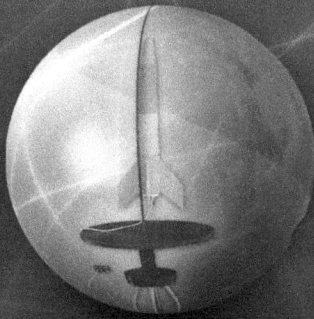

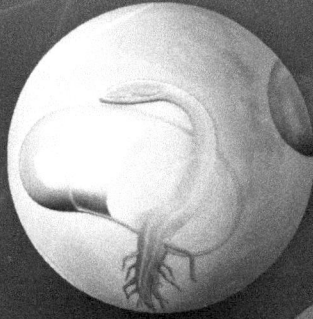

Rebecca W. Keller, PhD

REAL SCIENCE 4 Kids

Copyright © 2021 by Rebecca Woodbury, Ph.D.

All rights reserved. No part of this publication may be reproduced, stored in a retrieval system, or transmitted, in any form or by any means, electronic, mechanical, photocopying, recording, or otherwise, without prior written permission from the publisher. No part of this book may be used or reproduced in any manner whatsoever without written permission.

Super Simple Science Experiments Laboratory Notebook
ISBN 978-1-953542-03-8

Published by Gravitas Publications Inc.
www.gravitaspublications.com
www.realscience4kids.com

GRAVITAS
PUBLICATIONS

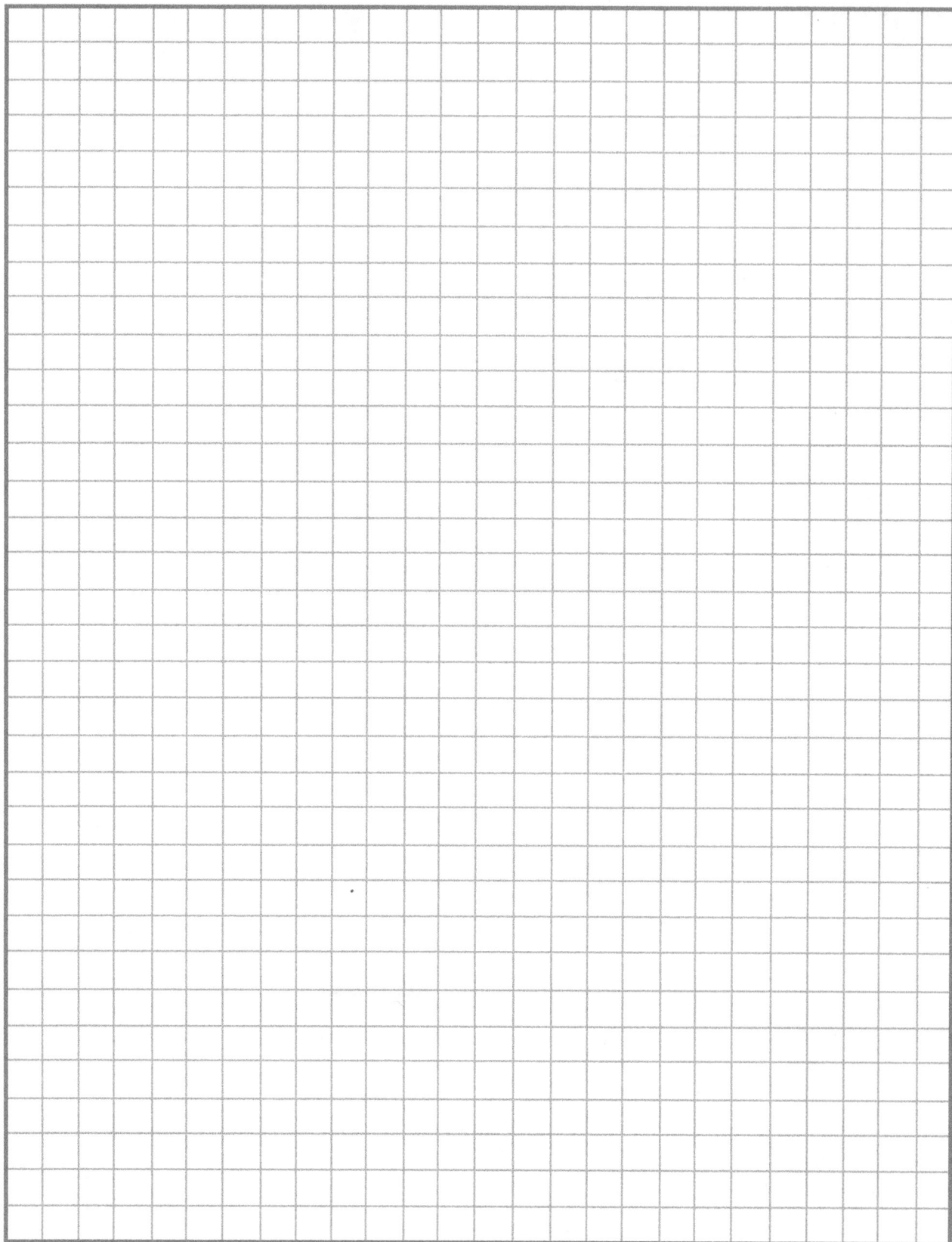

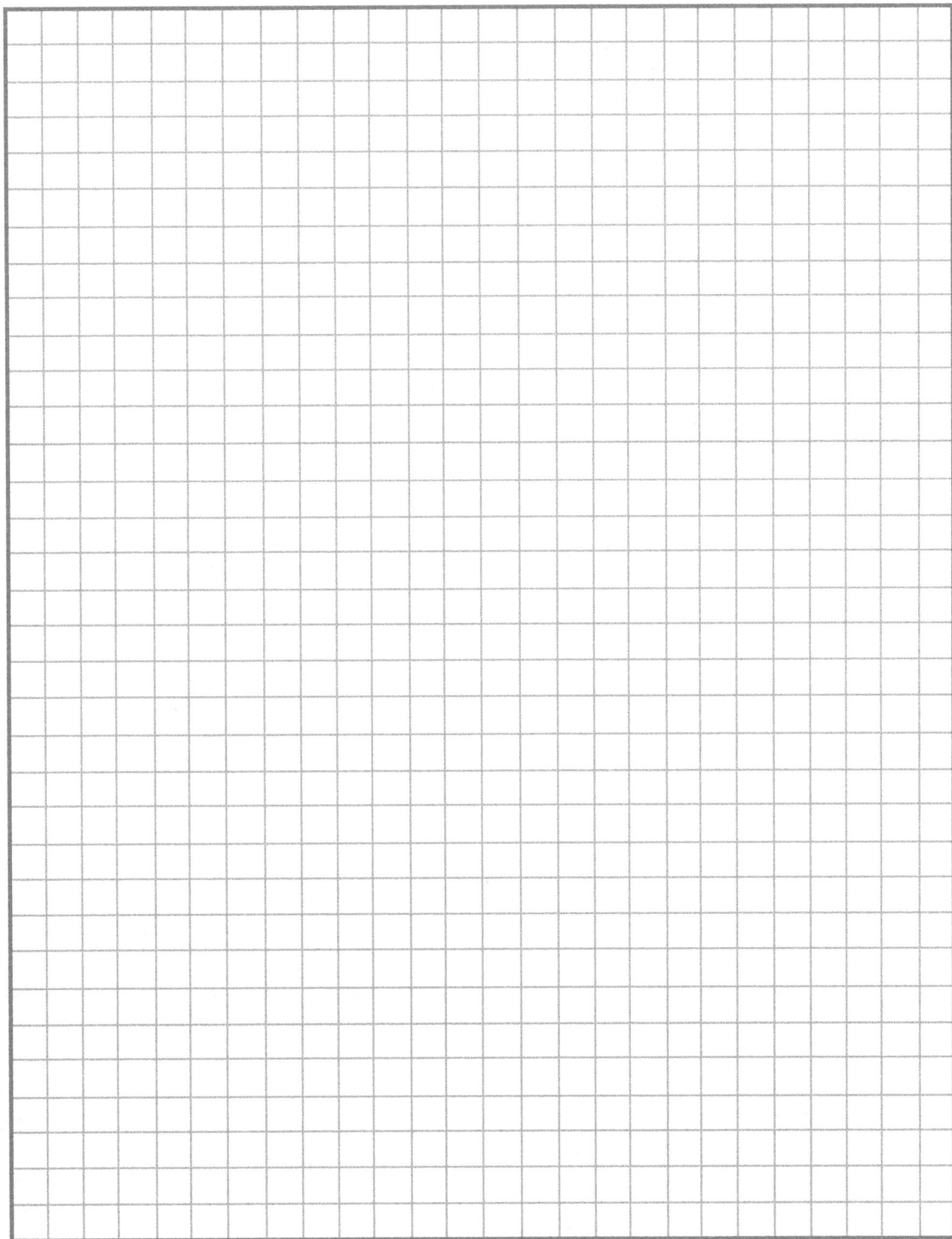

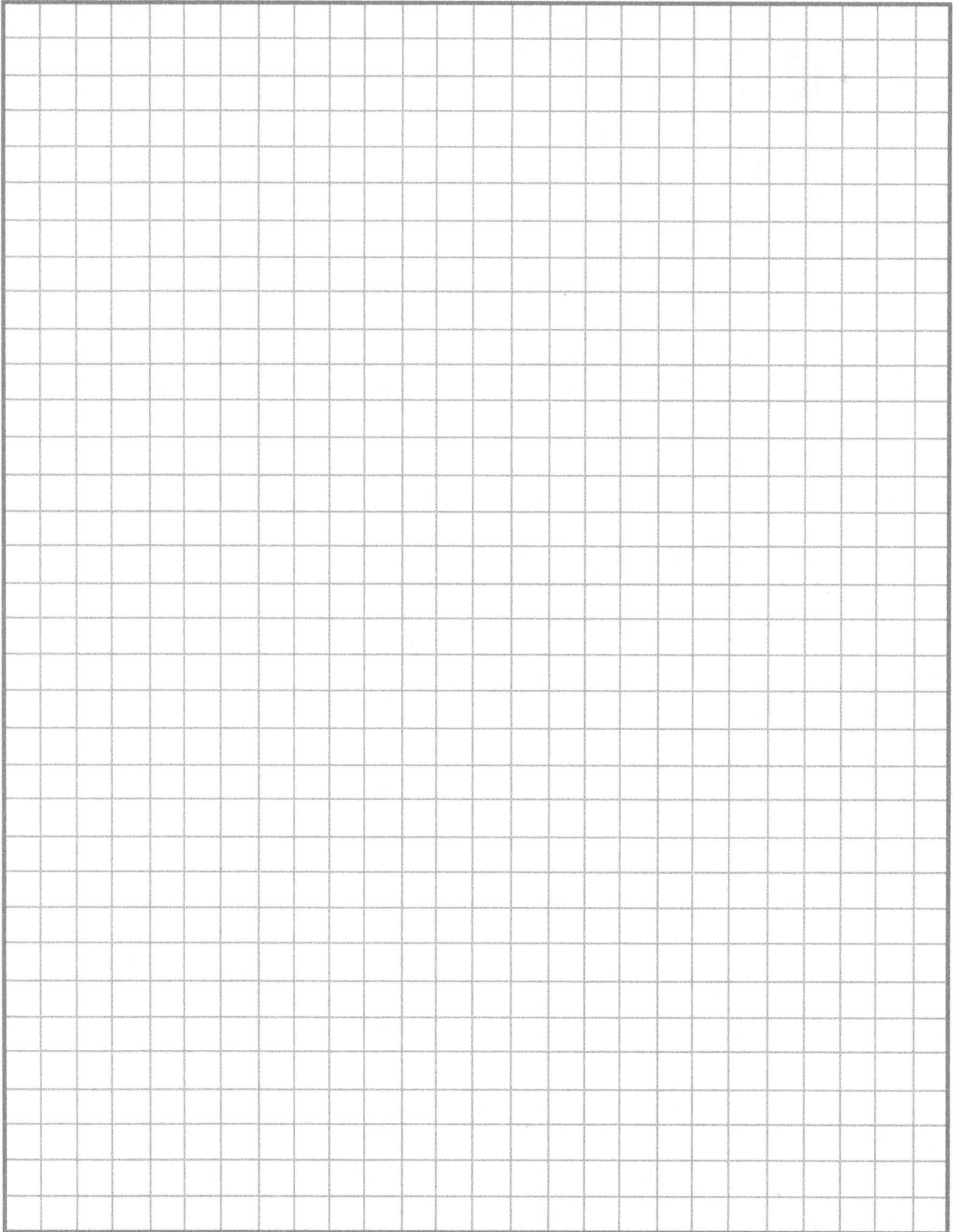

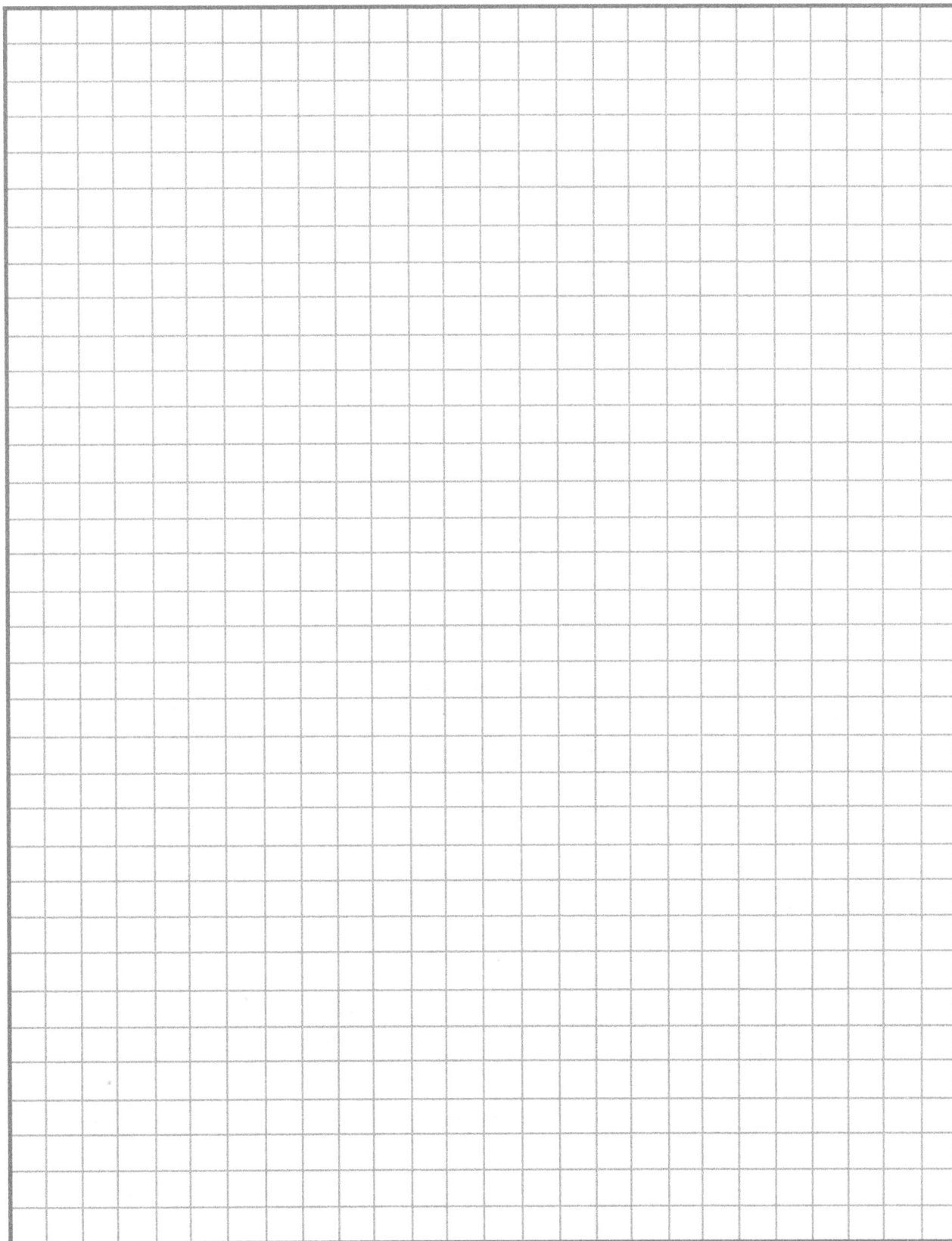

_____

_____

_____

_____

_____

_____

_____

_____

_____

_____

_____

_____

_____

_____

_____

_____

_____

_____

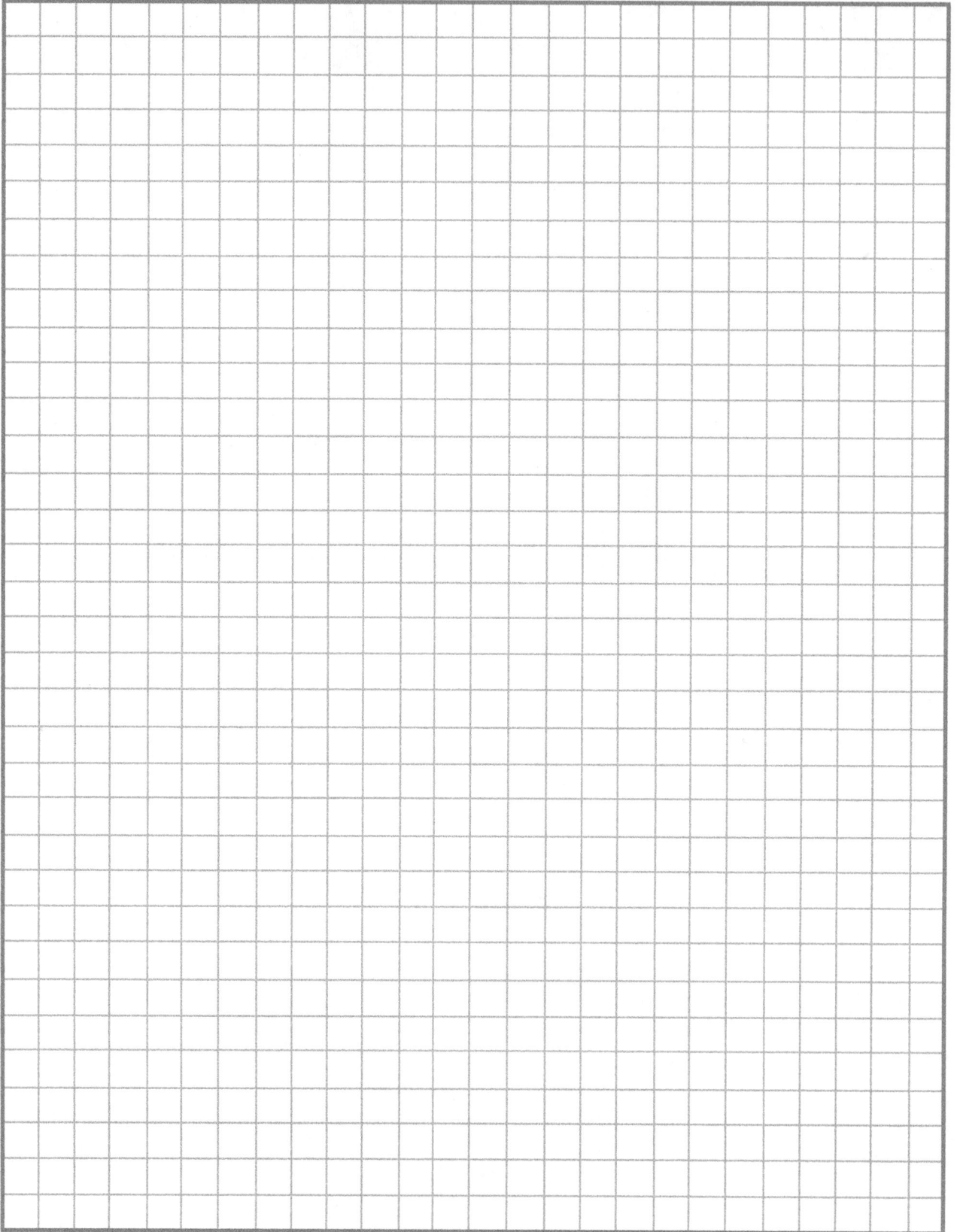

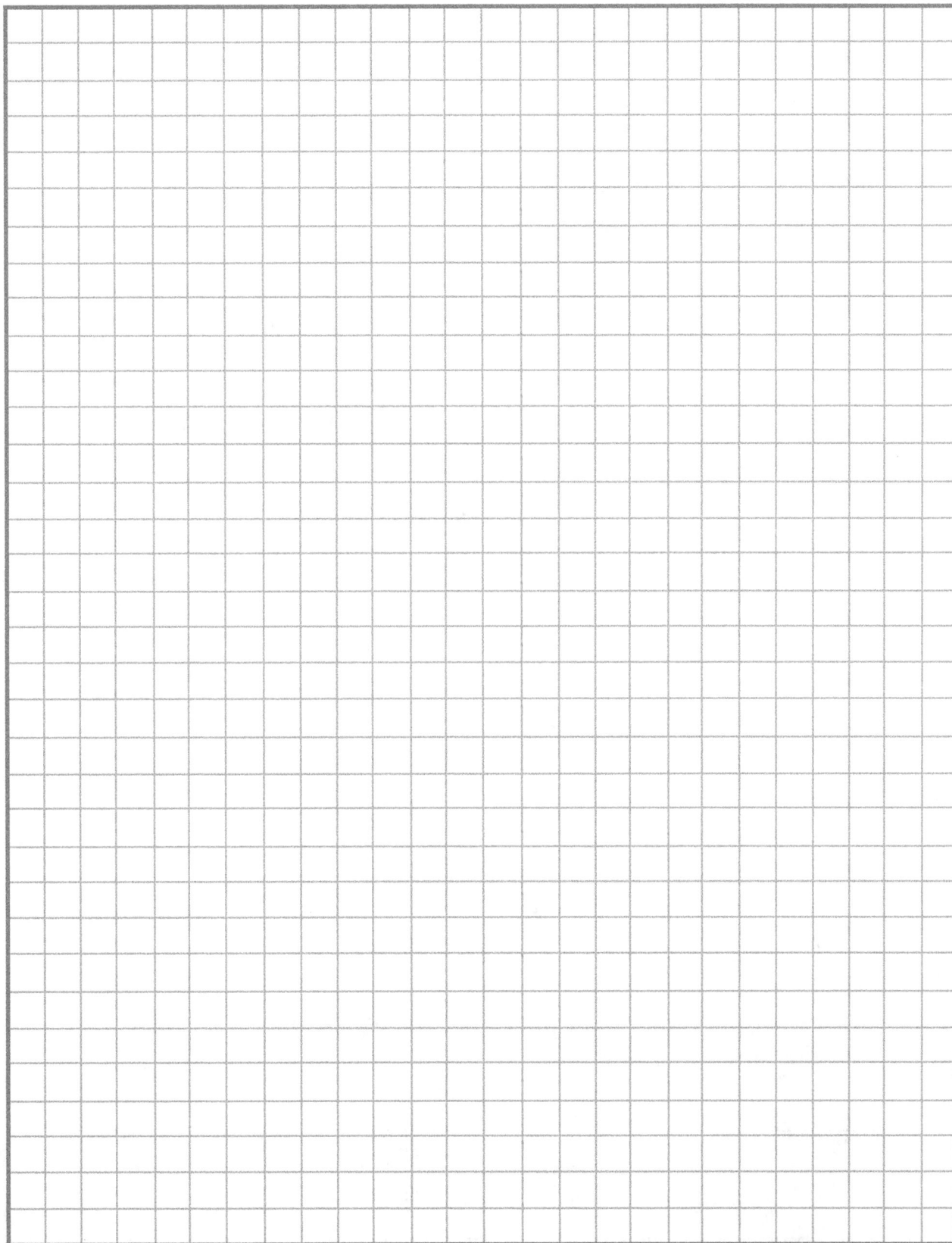

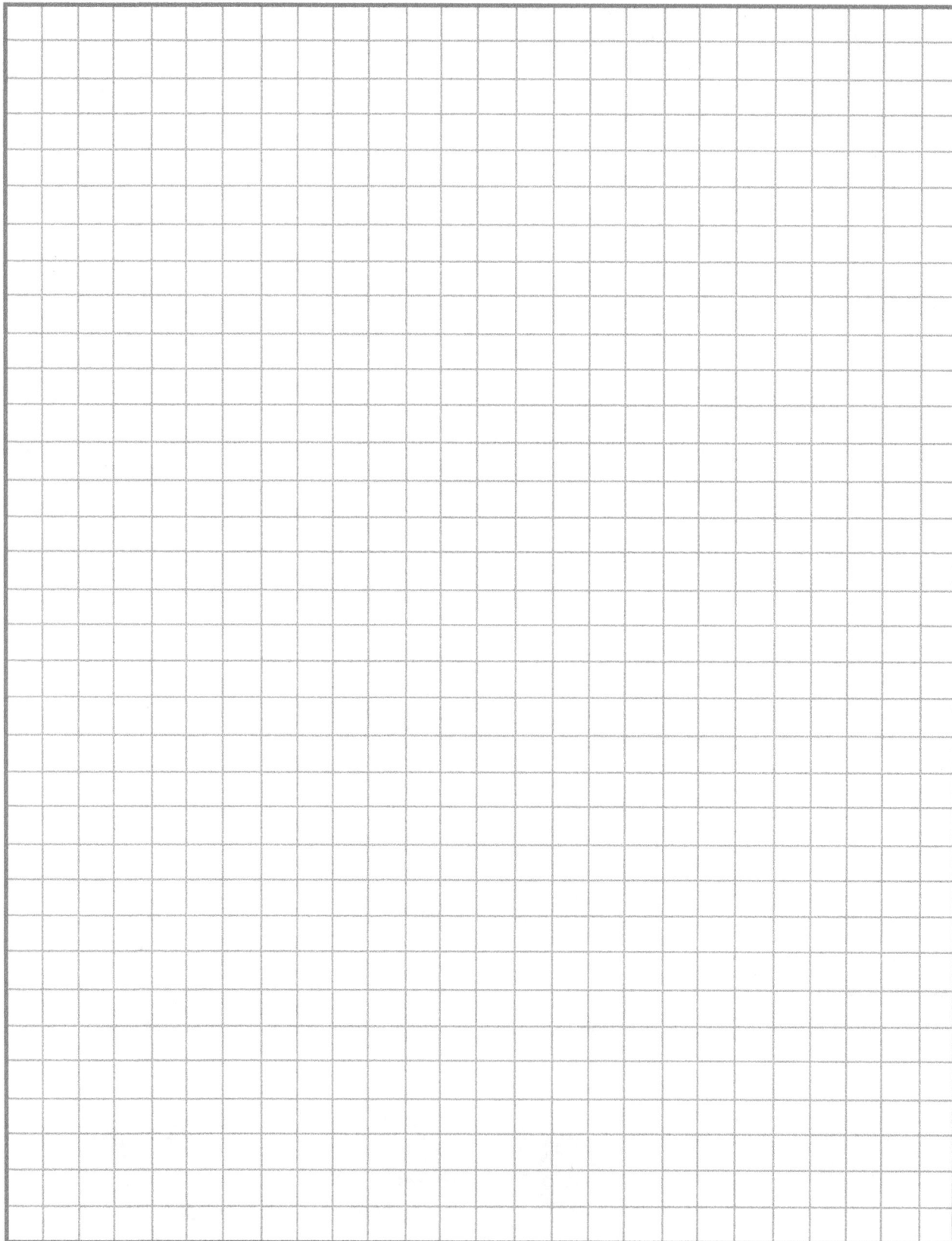

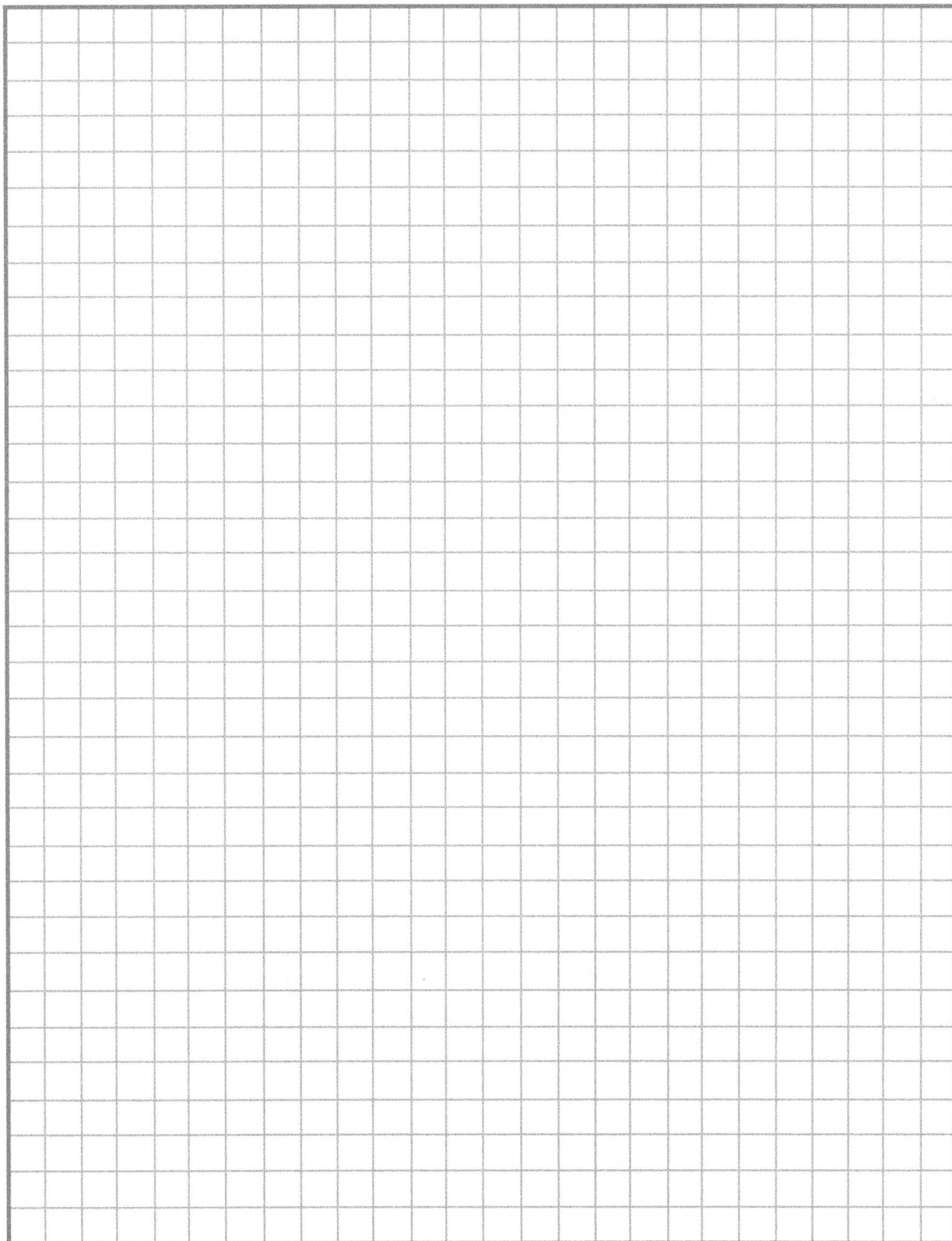

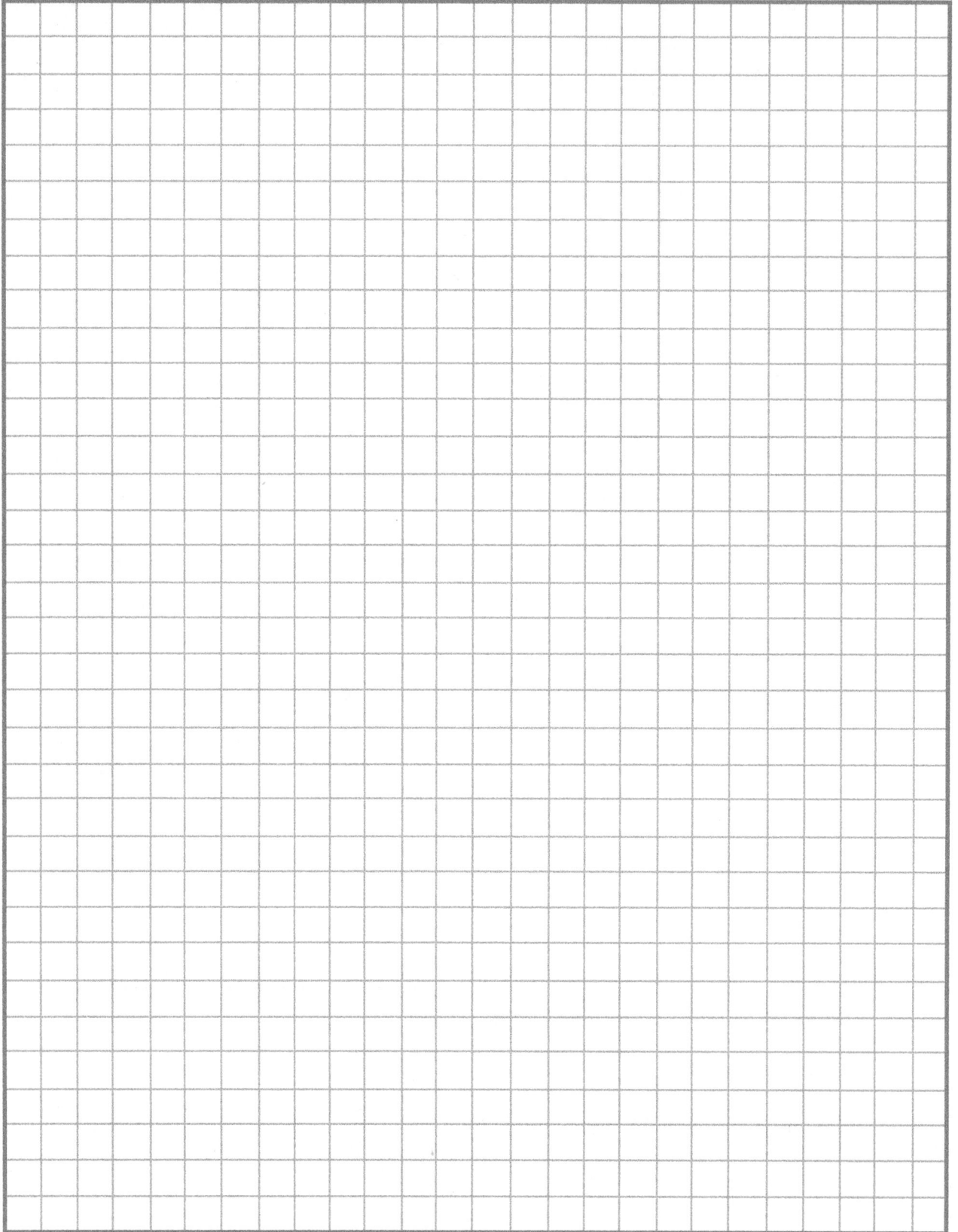

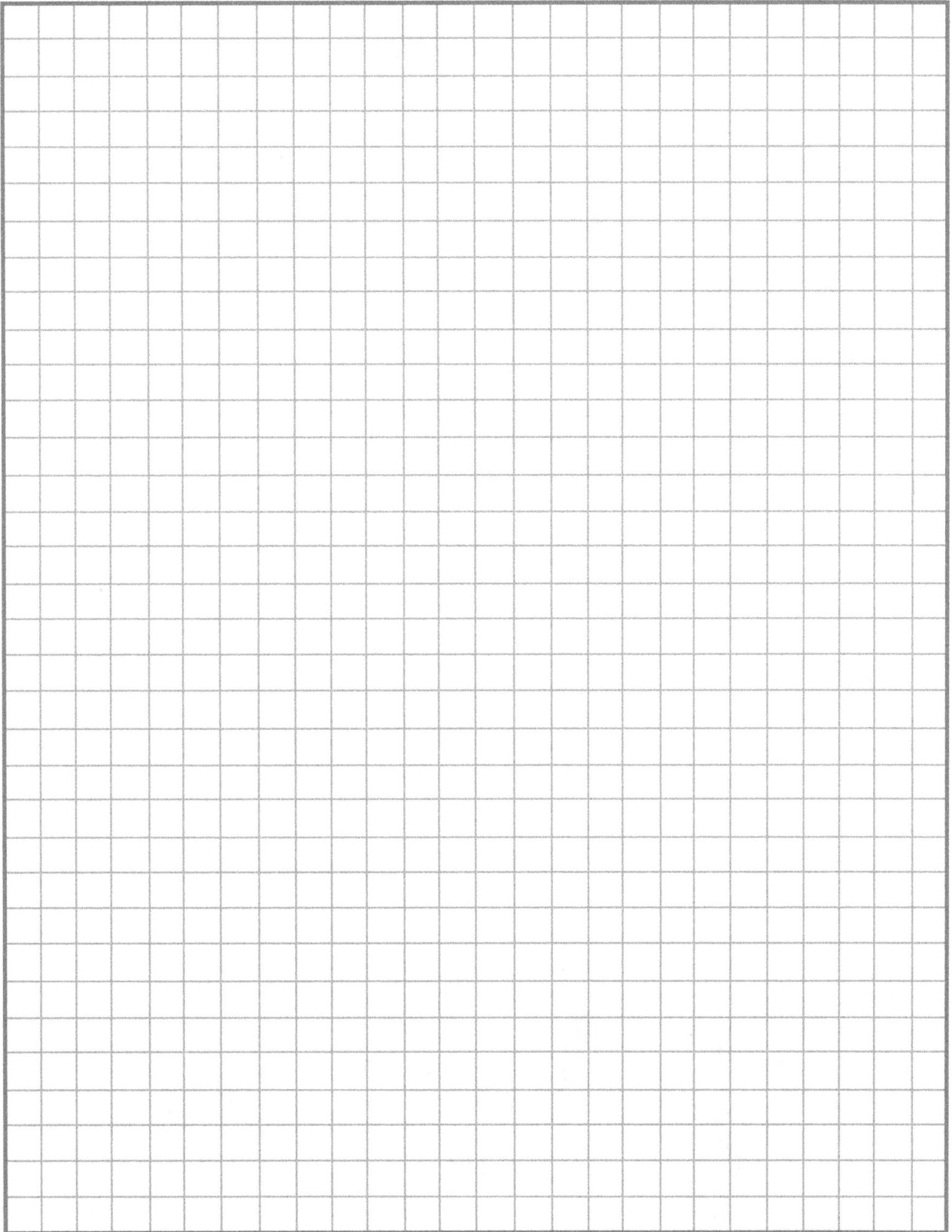

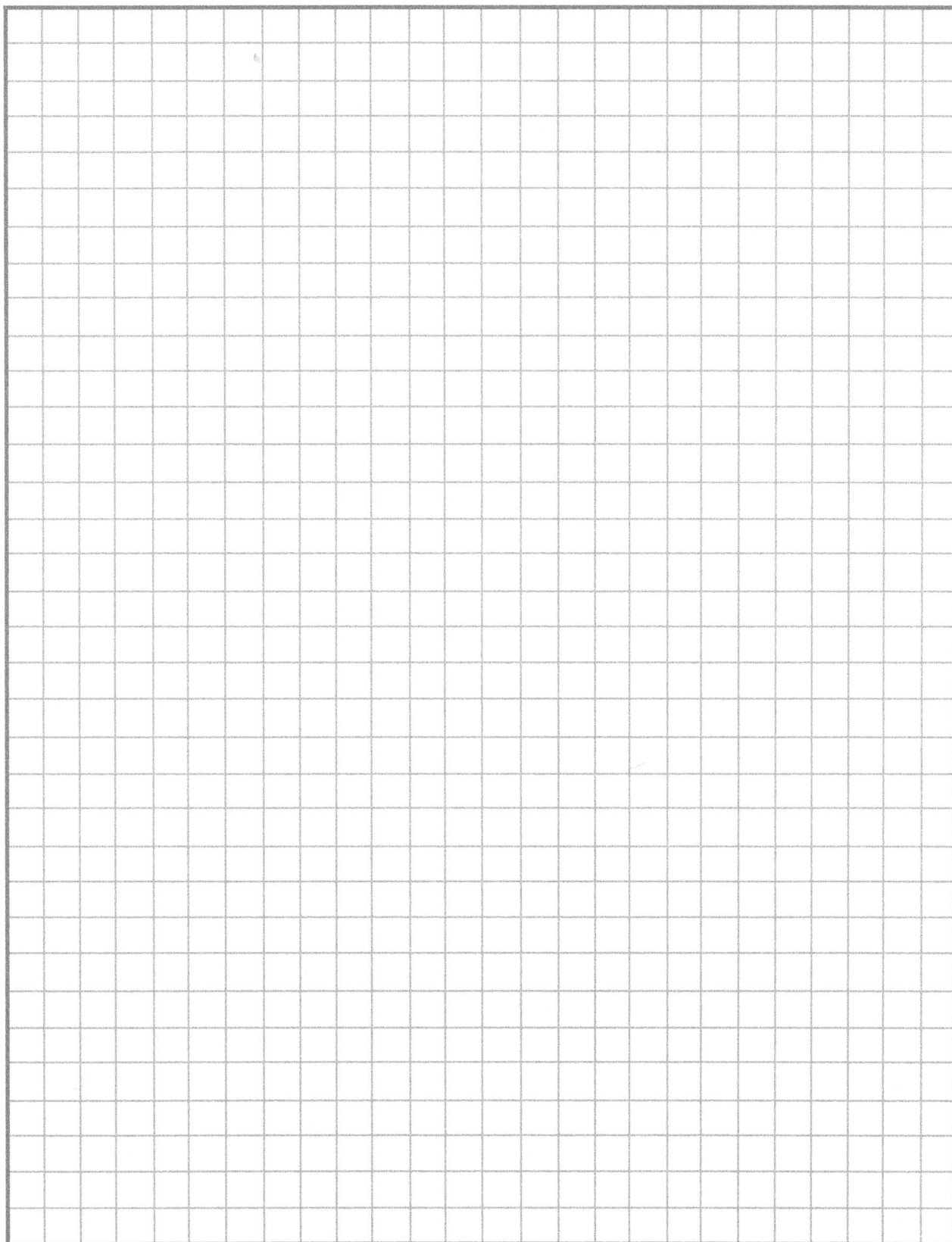

_____

_____

_____

_____

_____

_____

_____

_____

_____

_____

_____

_____

_____

_____

_____

_____

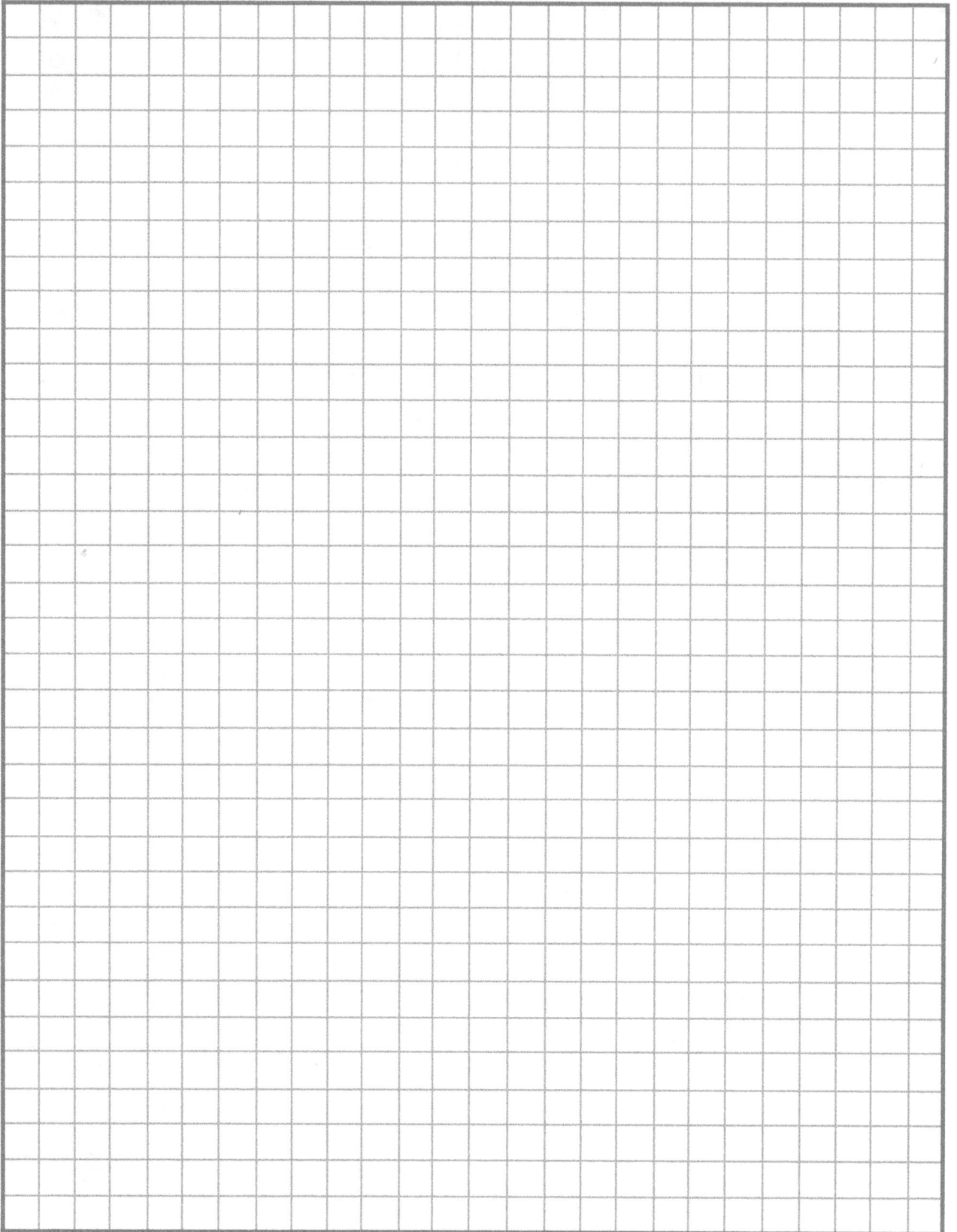

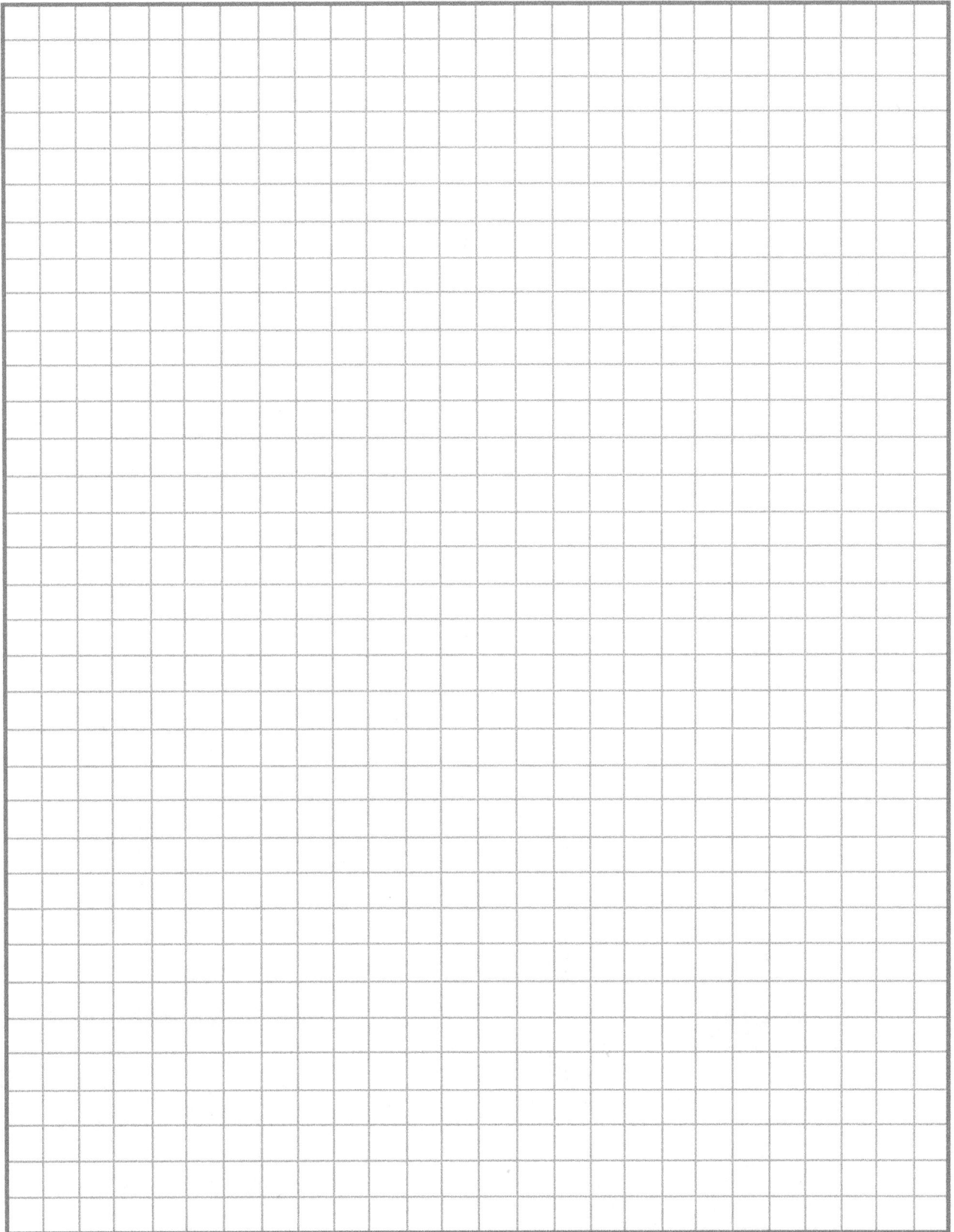

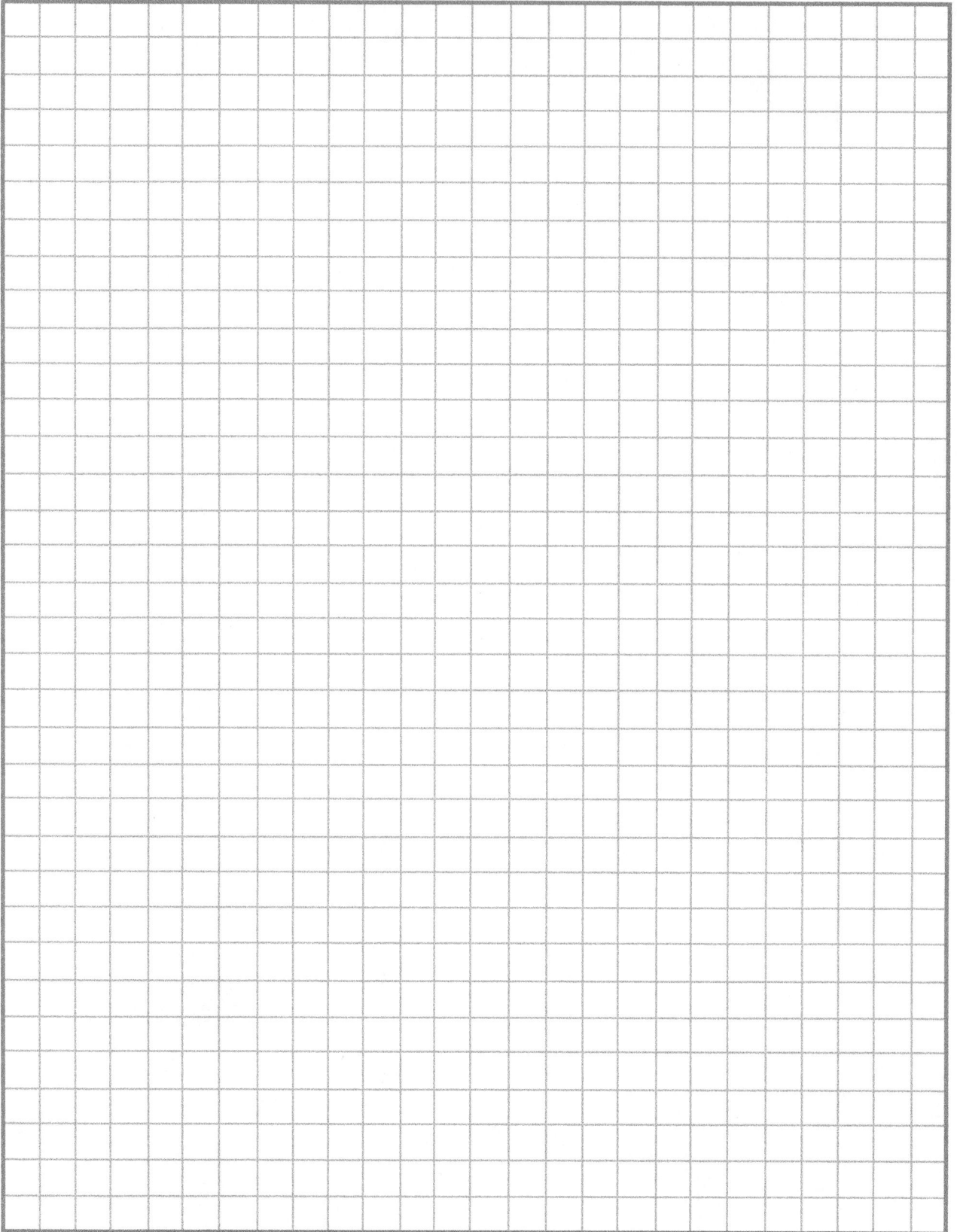

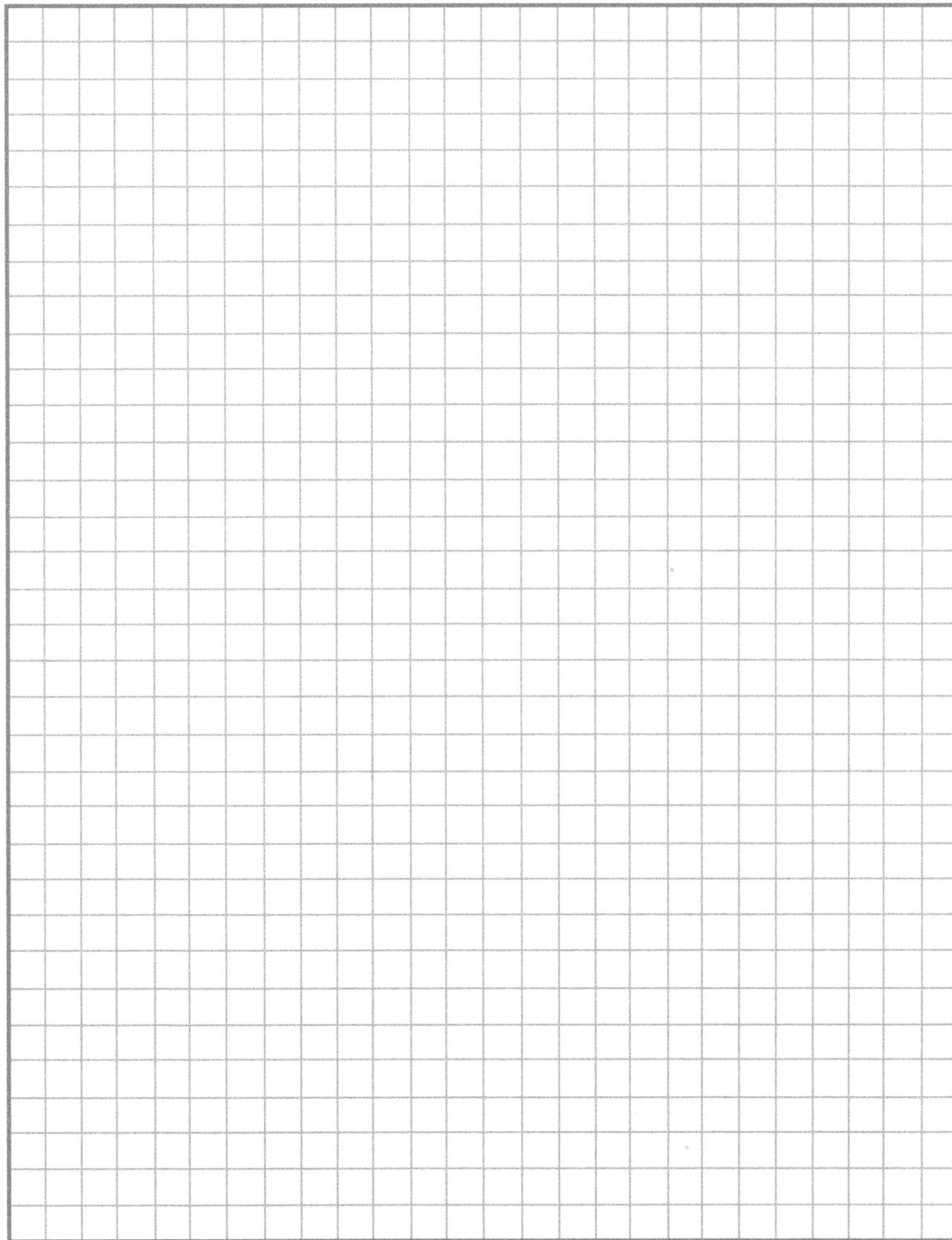

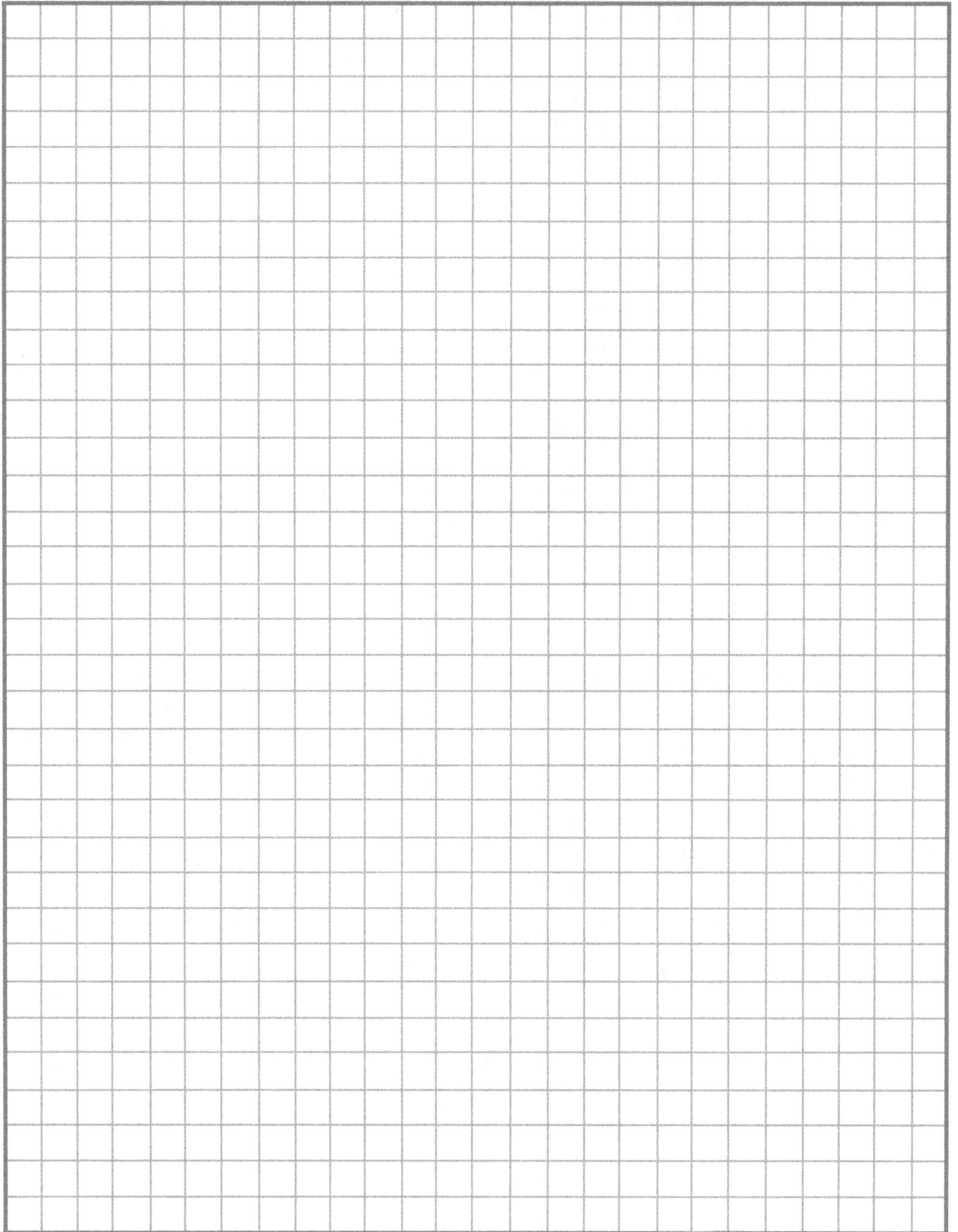

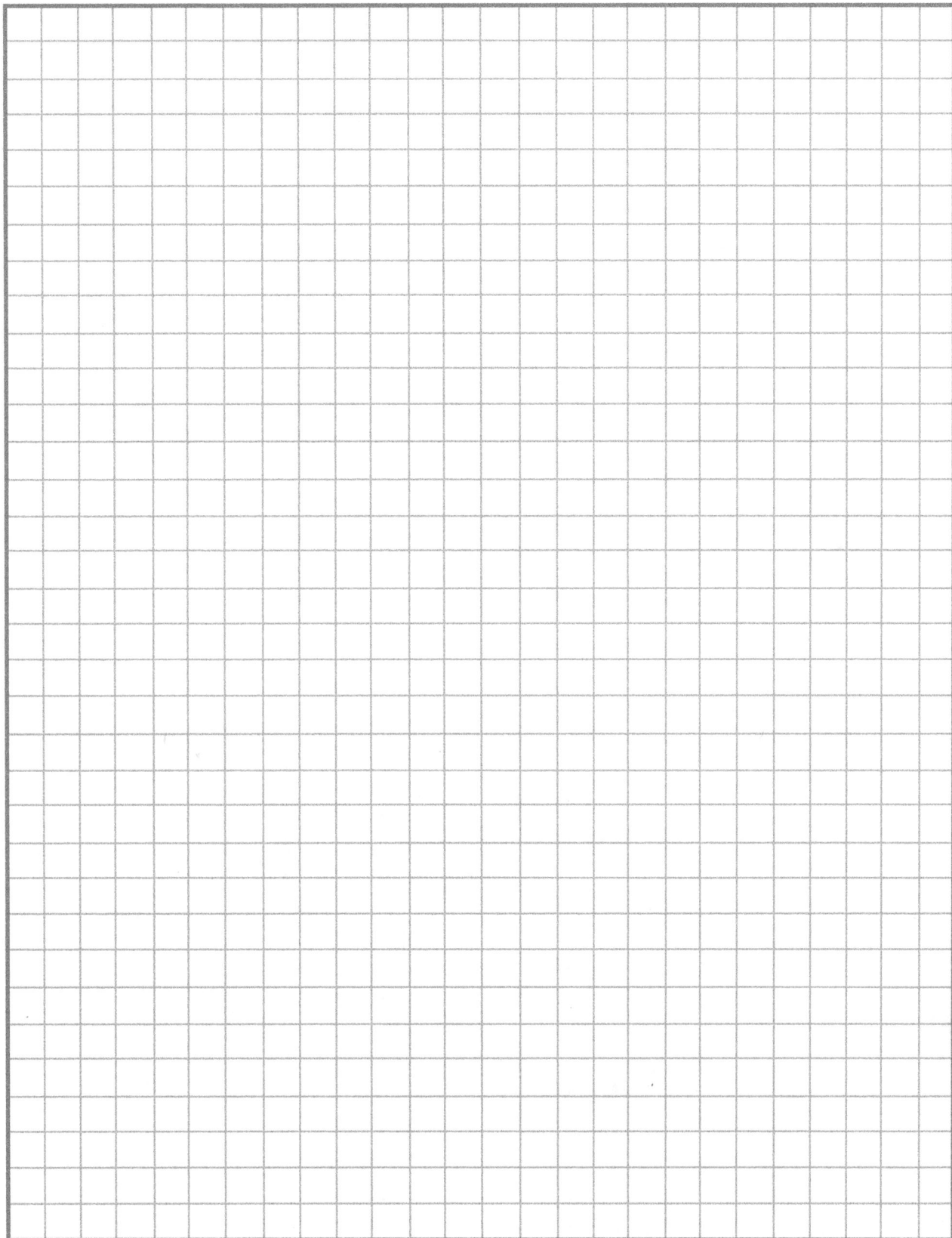

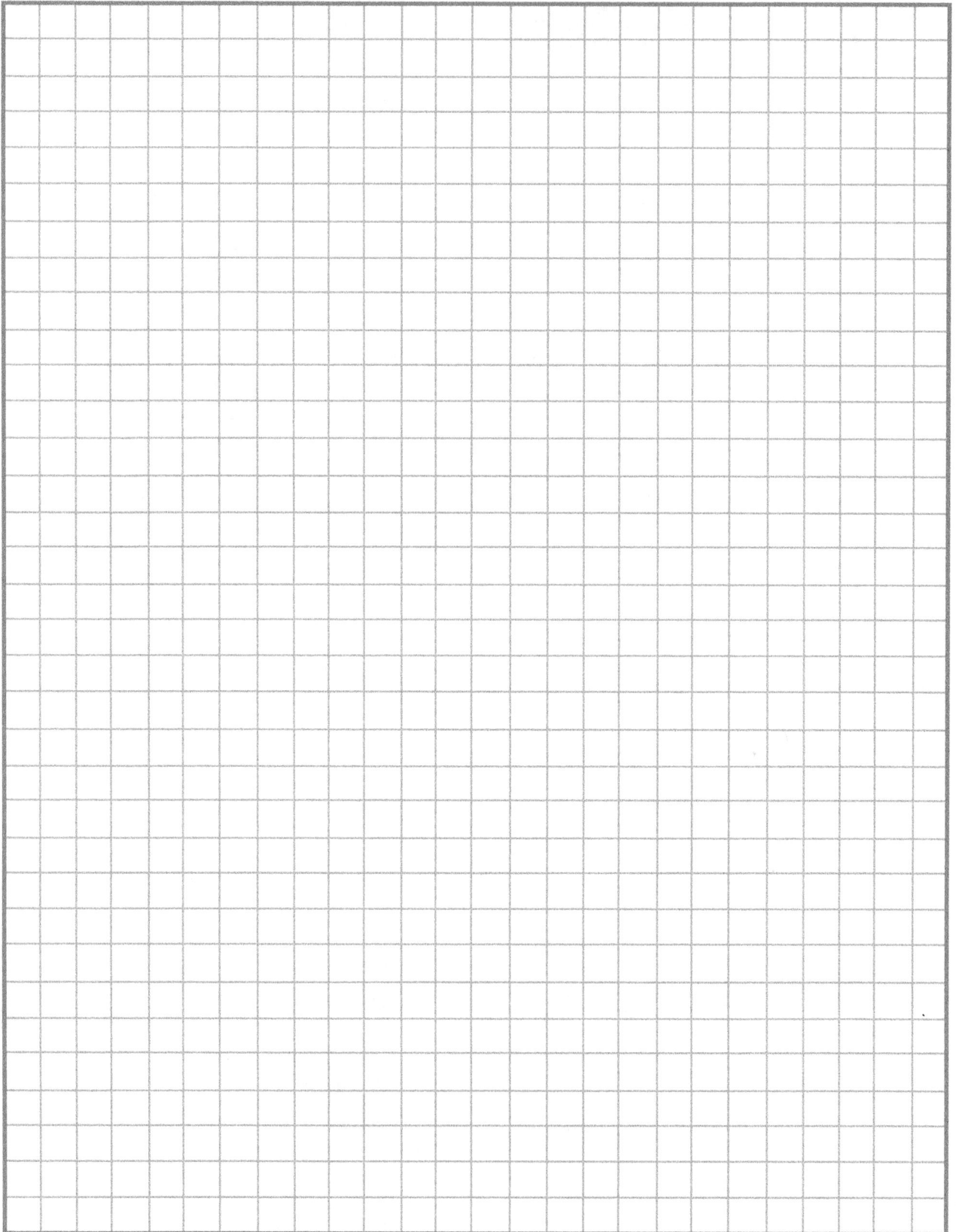

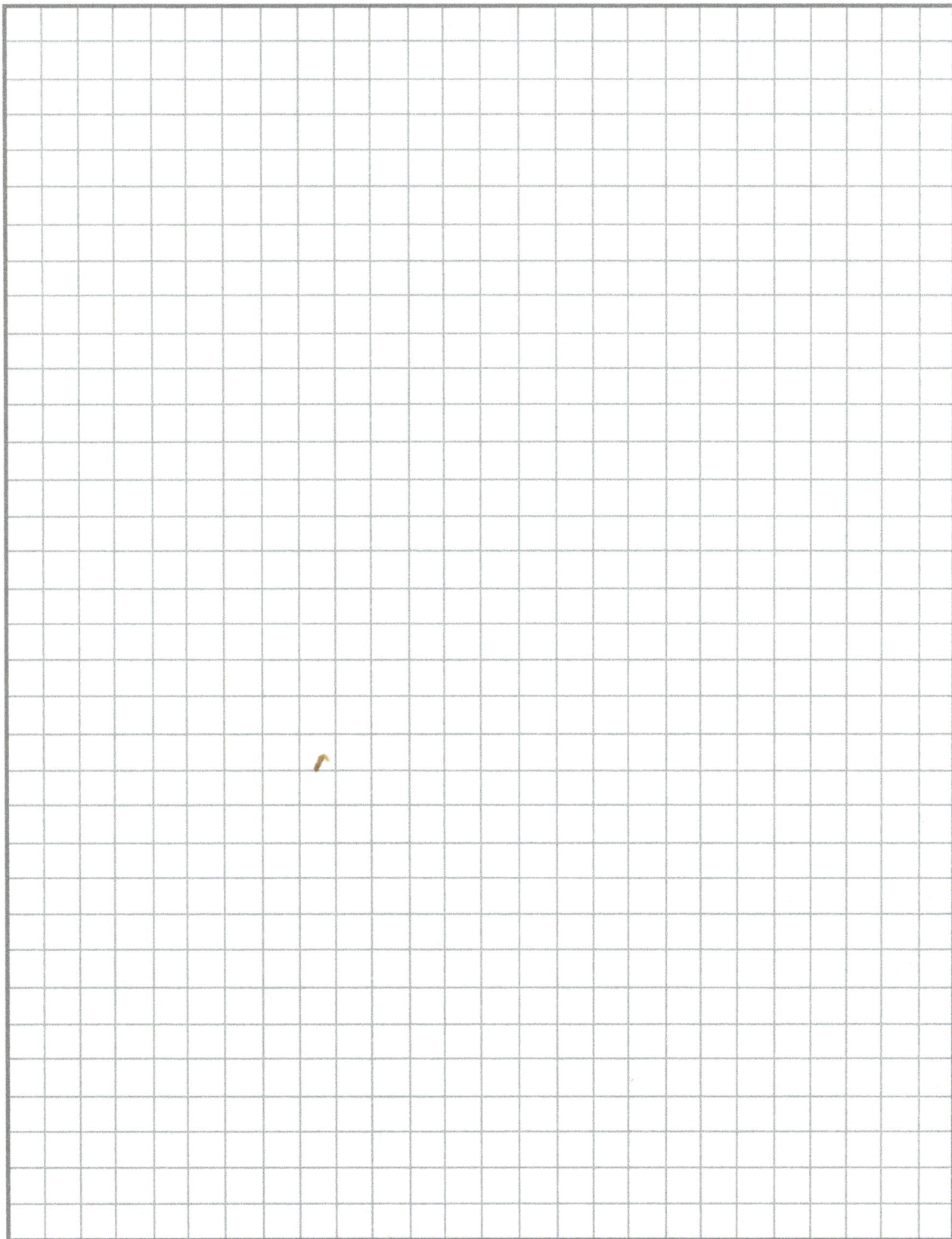

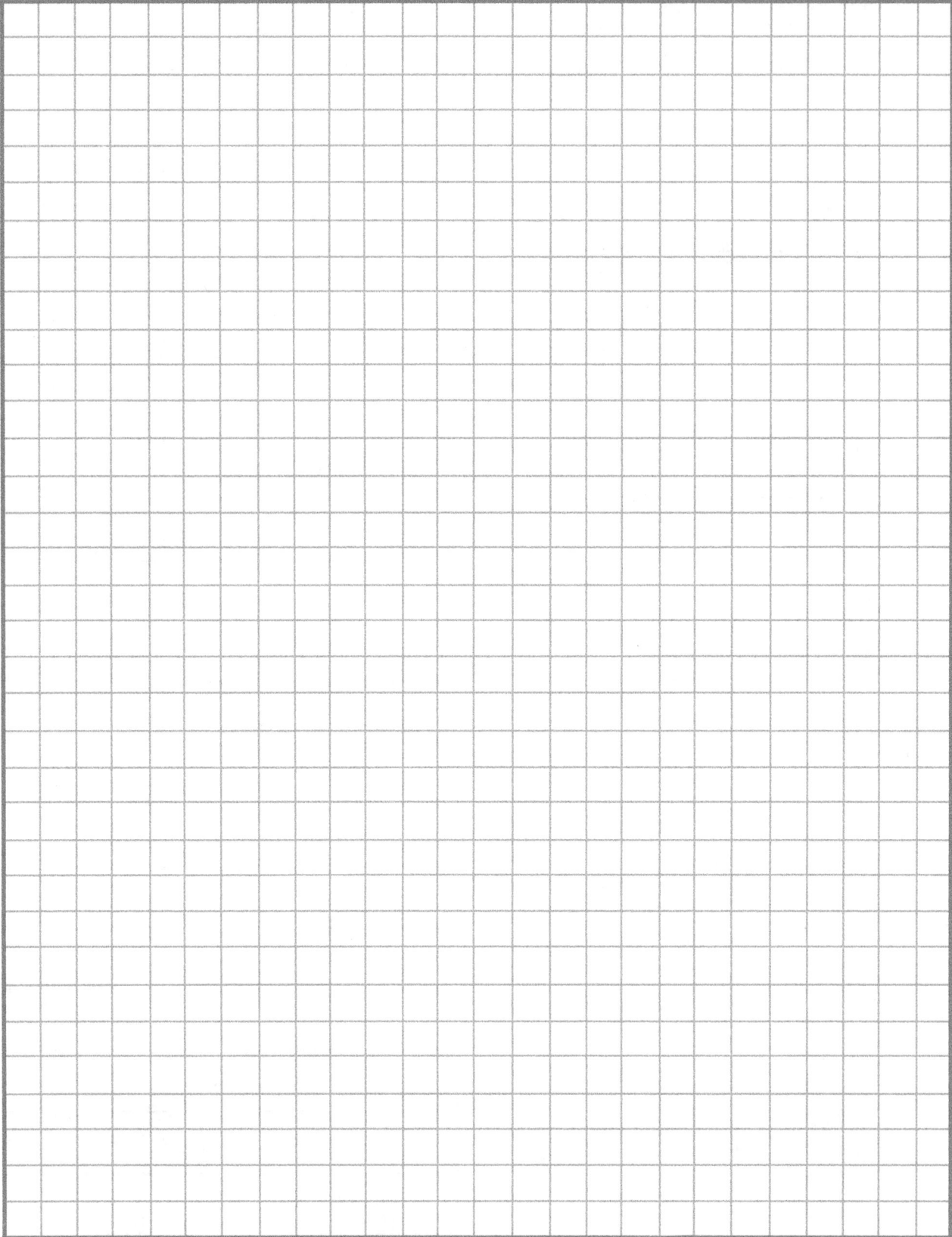

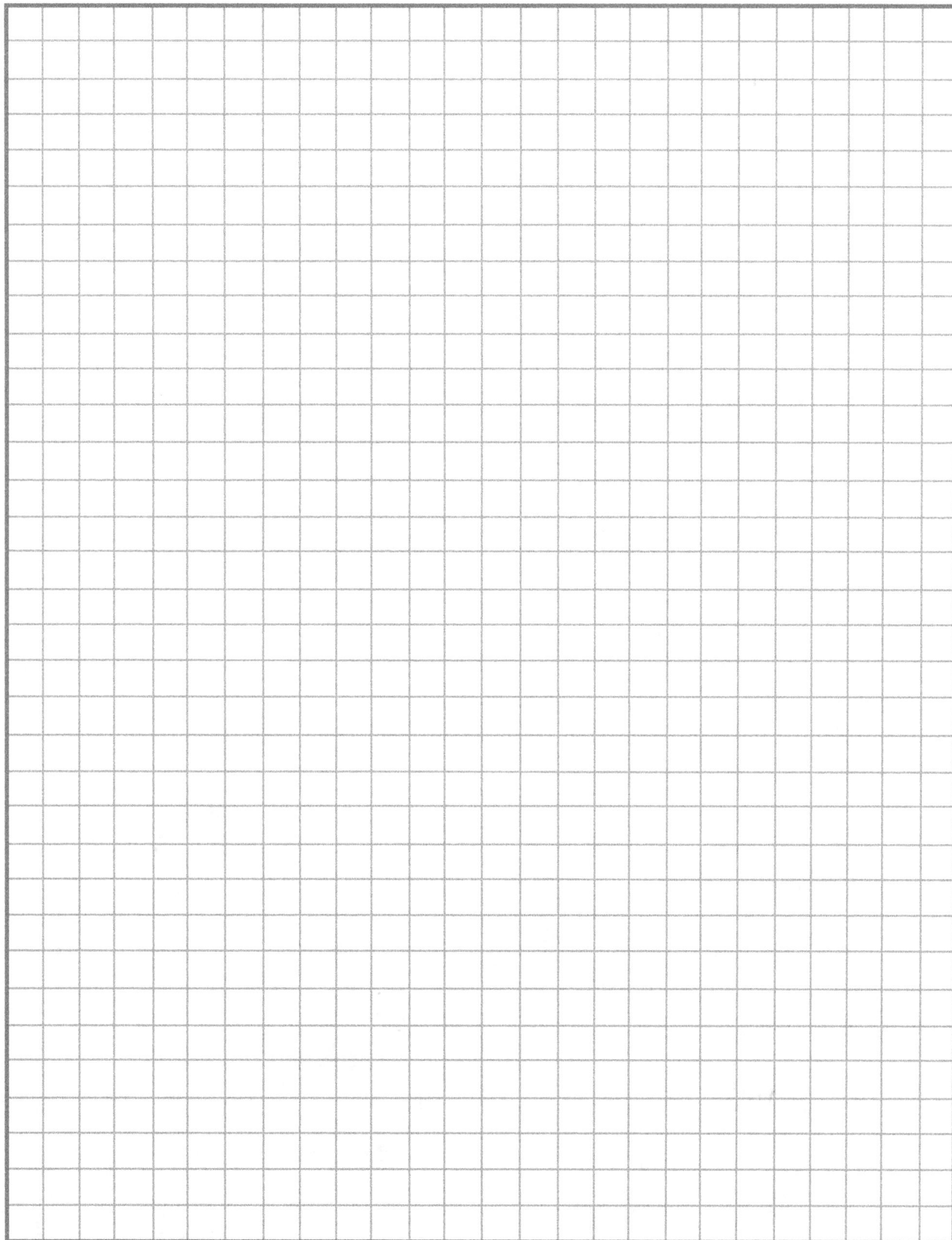

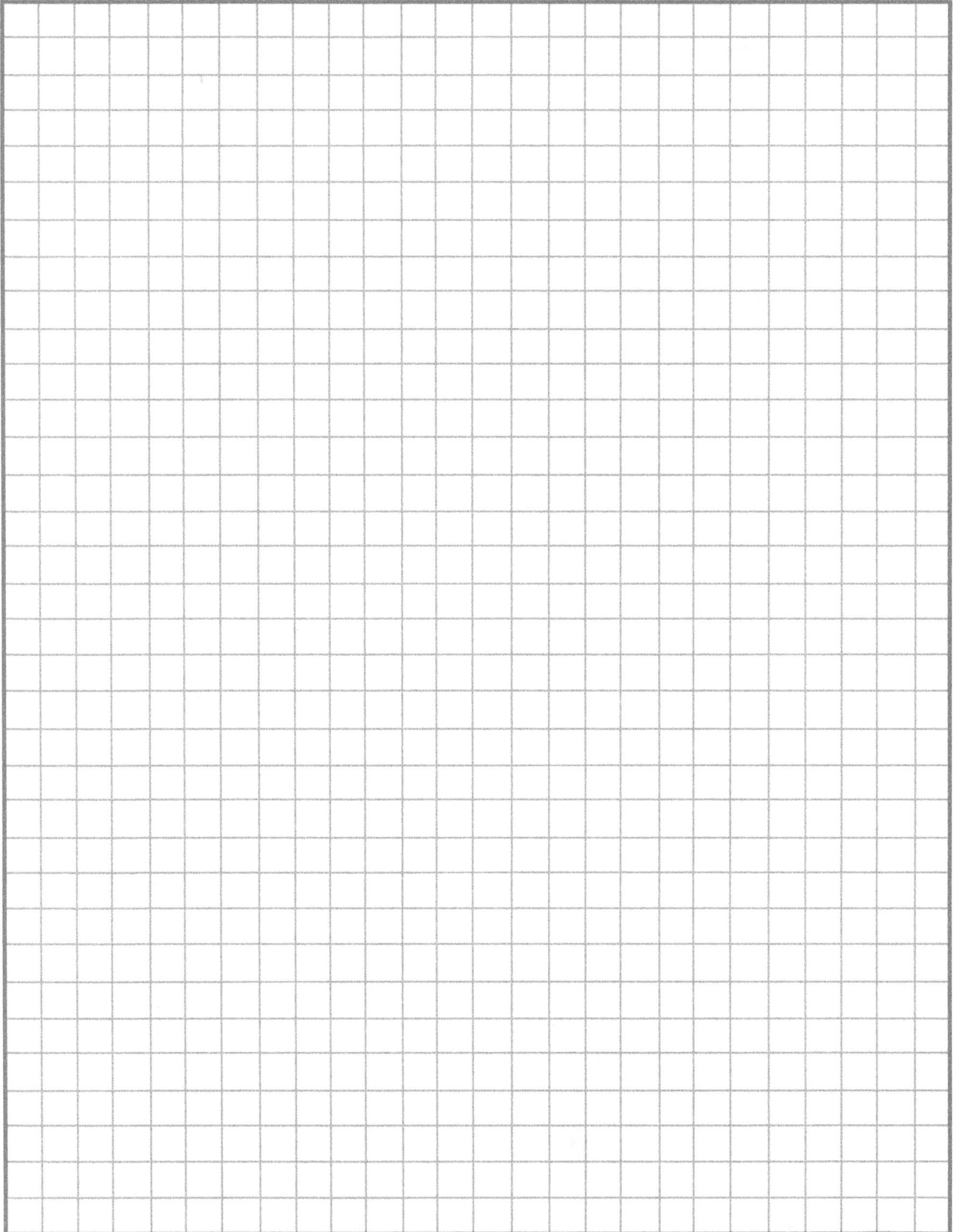

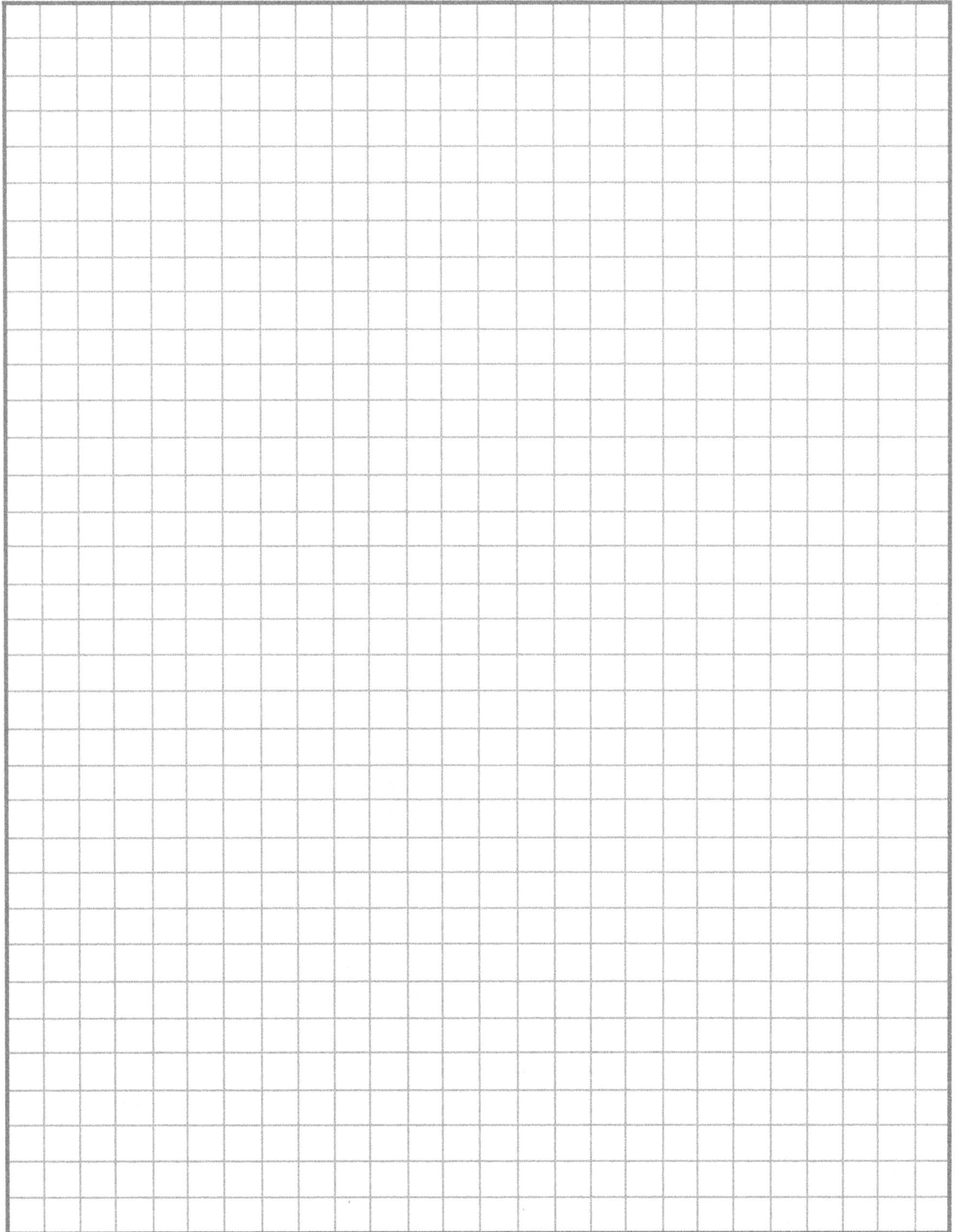

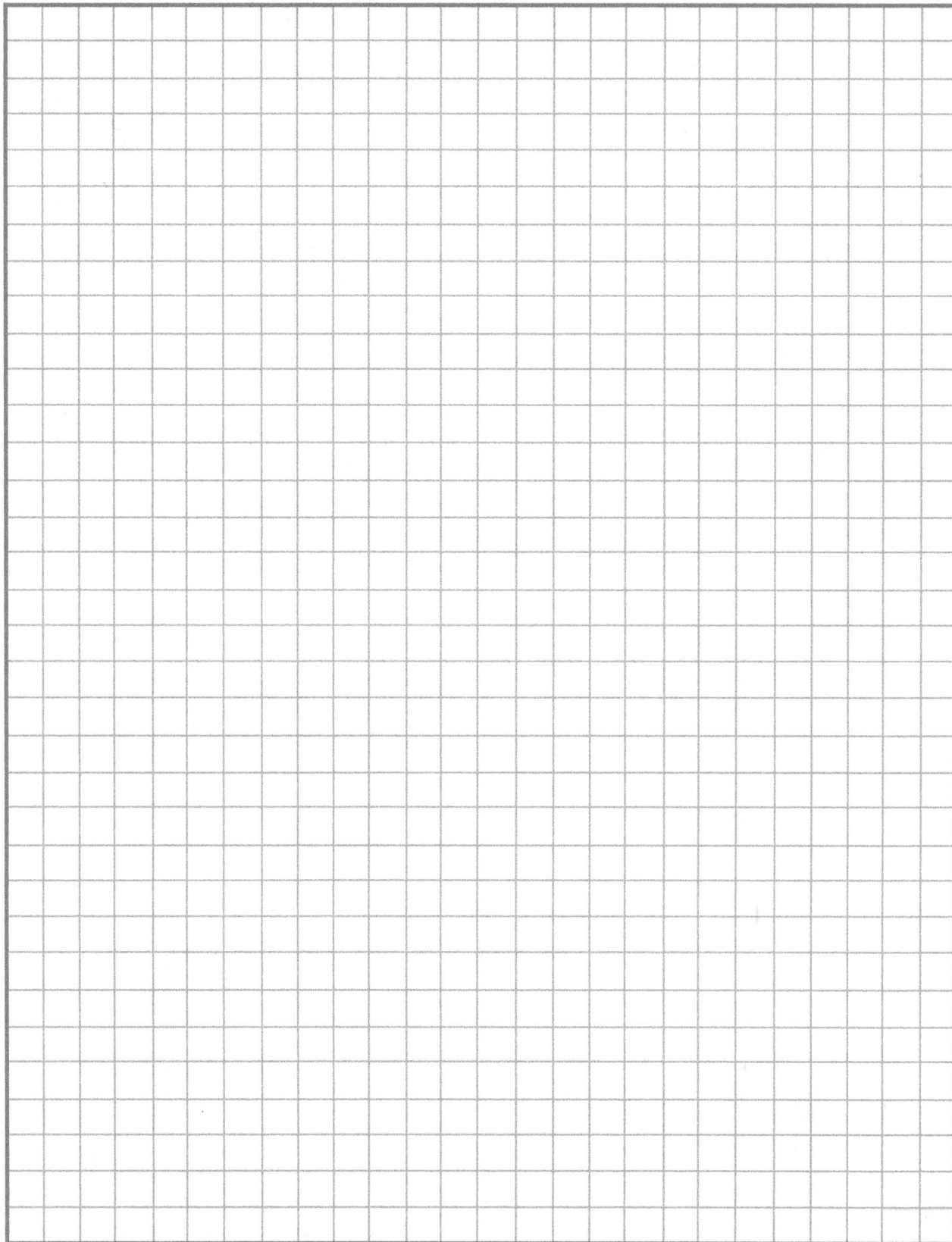

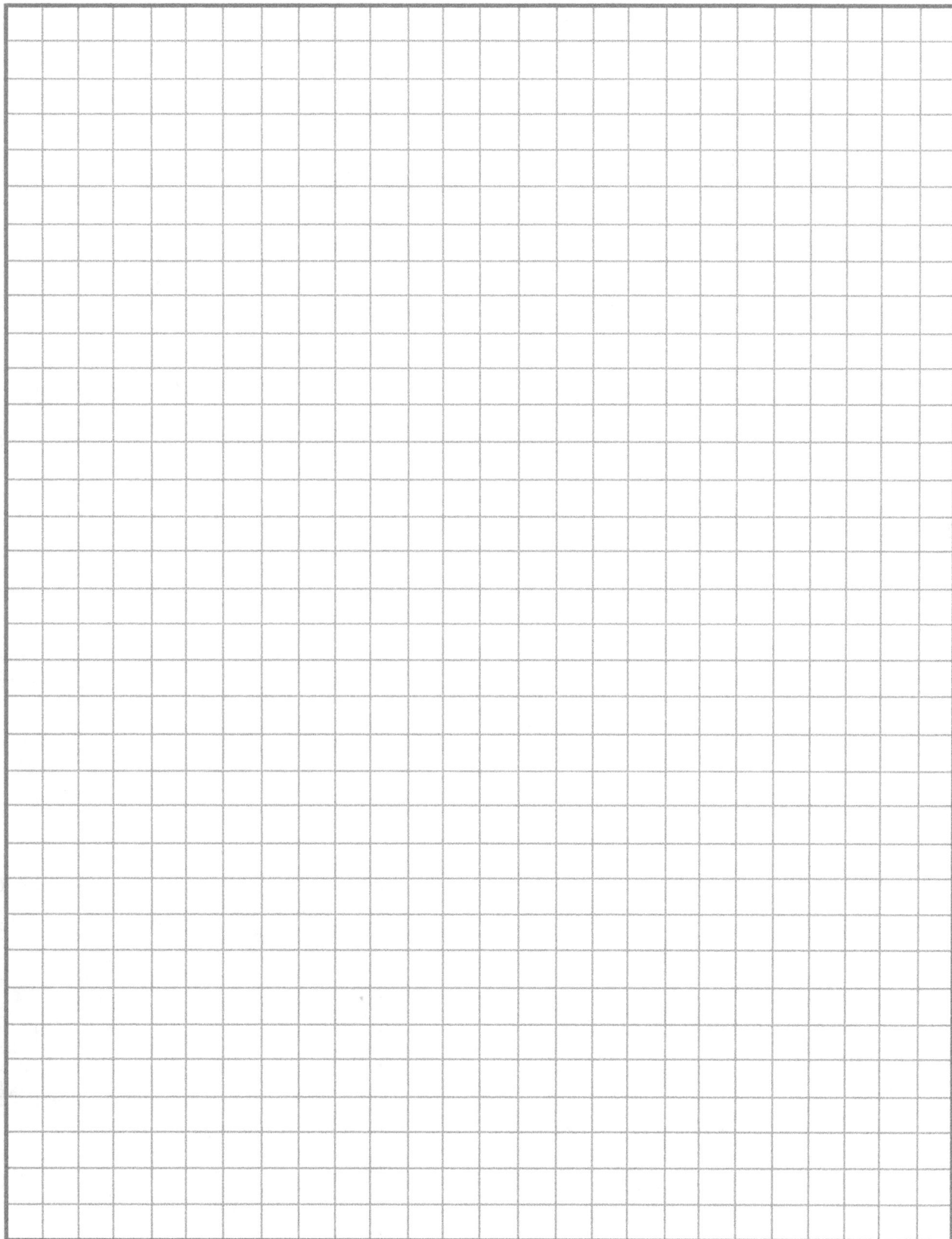

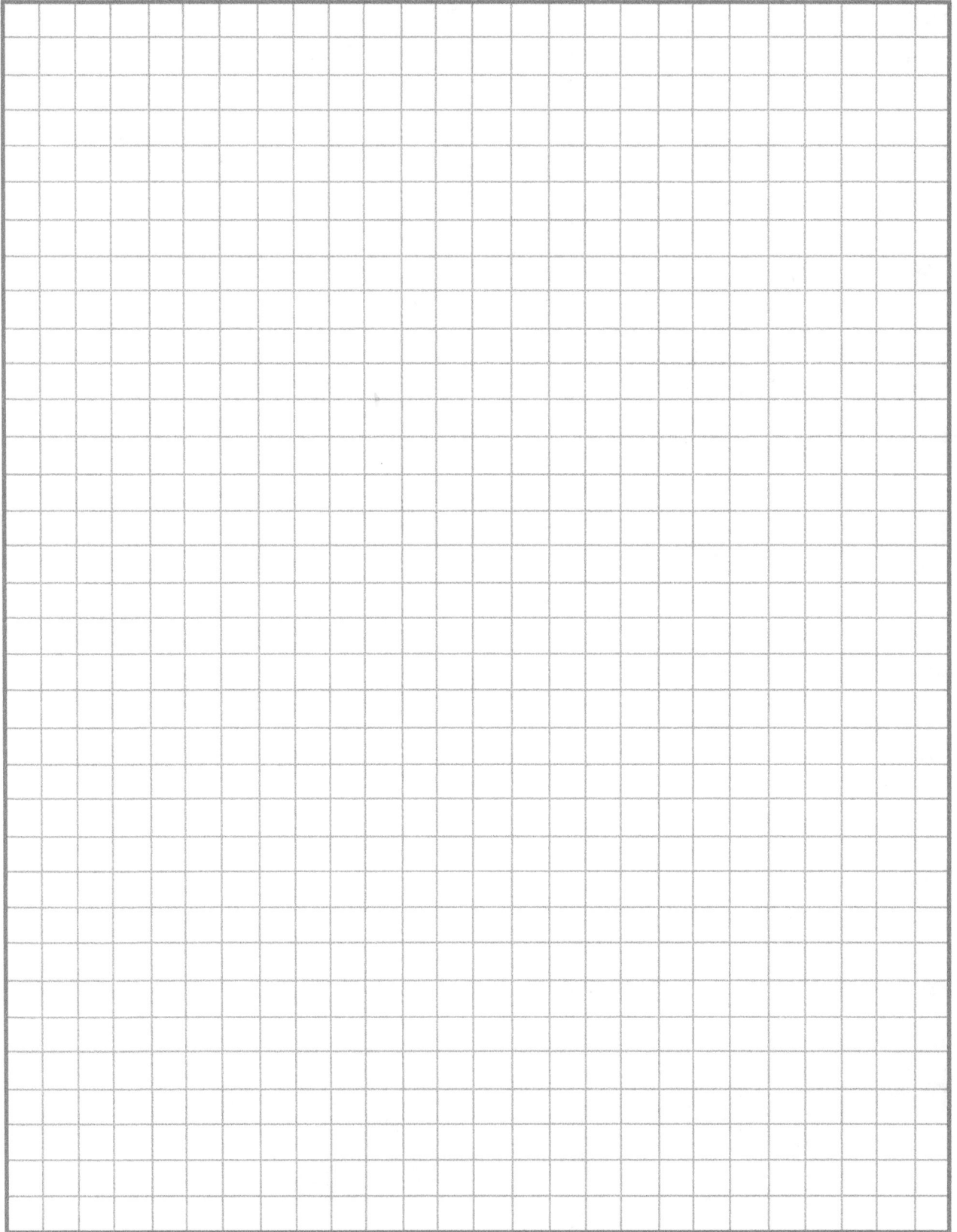

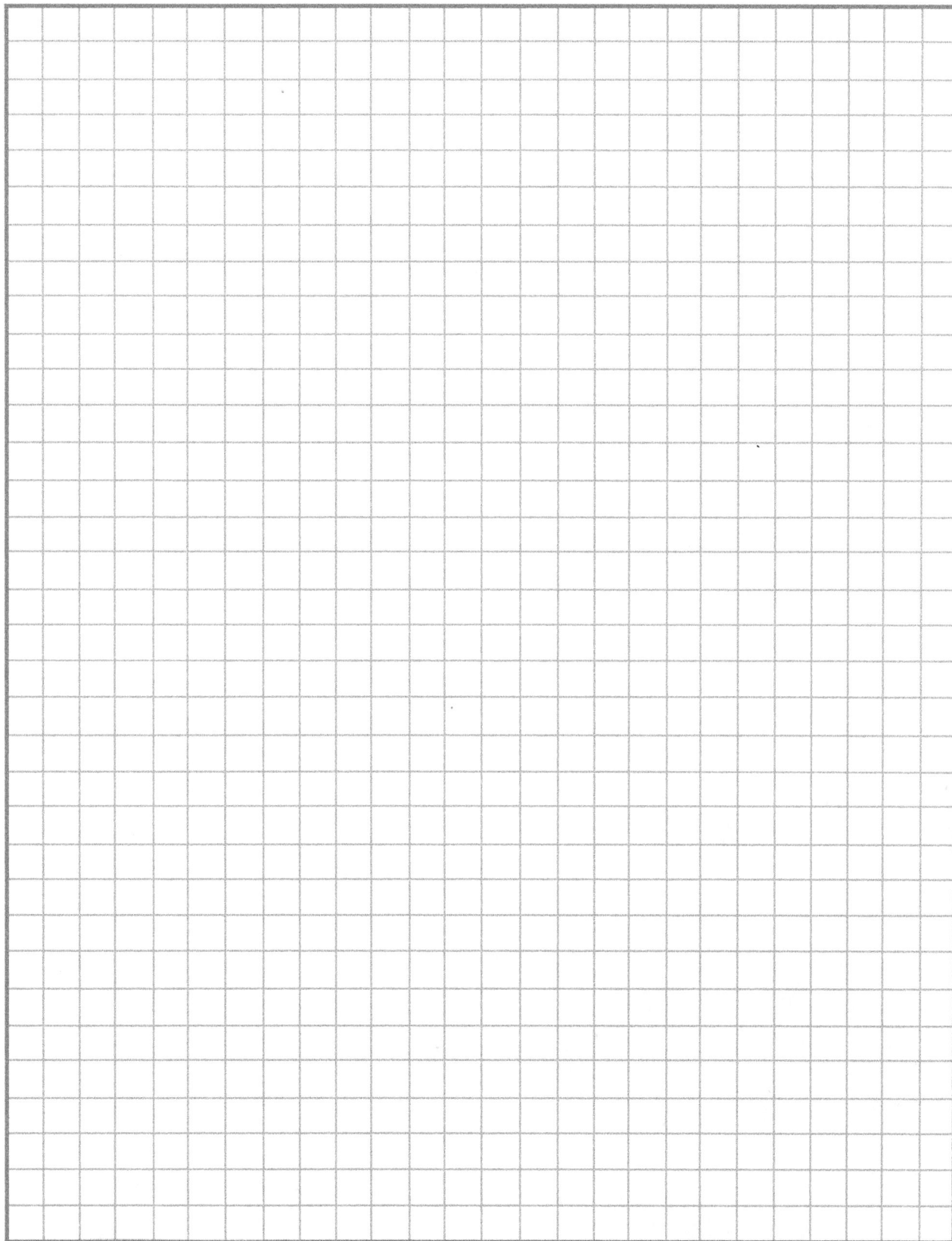

www.ingramcontent.com/pod-product-compliance
Lightning Source LLC
Chambersburg PA
CBHW062028210326
41519CB00060B/7201